CHICA'S TALES

CHICA'S TALES

Wild Animals in Santa Fe

by
CHICA WINDY RIDGE
with
JAMES BAKER

SUNSTONE PRESS
SANTA FE

Sunstone books may be purchased for educational, business, or sales promotional use.
For information please write: Special Markets Department, Sunstone Press,
P.O. Box 2321, Santa Fe, New Mexico 87504-2321.
Printed on acid-free paper
∞

———————————

Library of Congress Cataloging-in-Publication Data

Names: Baker, James, 1939- author.
Title: Chica's tales : some wild animals in Santa Fe / by Chica Windy Ridge
    with James Baker.
Description: Santa Fe : Sunstone Press, [2024] | Audience: Ages 9 |
    Audience: Grades 4-6 | Summary: "A dog named Chica narrates a photo
    story of wild animals in their natural habitat in the foothills of Santa
    Fe, New Mexico"-- Provided by publisher.
Identifiers: LCCN 2023049447 | ISBN 9781632936318 (paperback) | ISBN
    9781632936325 (hardcover)
Subjects: LCSH: Chica Windy Ridge. | Mountain animals--New Mexico--Santa
    Fe--Juvenile literature. | Wildlife watching--New Mexico--Santa
    Fe--Juvenile literature.
Classification: LCC QL113 .B35 2024 | DDC 591.70978956--dc23/eng/20231115
LC record available at https://lccn.loc.gov/2023049447

———————————

WWW.SUNSTONEPRESS.COM
SUNSTONE PRESS / POST OFFICE BOX 2321 / SANTA FE, NM 87504-2321 /USA
(505) 988-4418

DEDICATION

This is for Suzanne, her children James and Meredith, and my children Russell and Lane.

—James Baker

PREFACE

Have you ever heard coyotes barking in the night? Or seen deer crossing the road expecting the traffic to stop? Or watched a group of crows chasing a hawk out of the area? In the foothills of Santa Fe we live in an ecology that includes wild animals. Some are seen, some are heard, and some leave evidence such as tracks in the snow and mud.

Who are these animals; how often do they come around; who lives among us; and who might just be passing through? A trail camera which could take photographs in both day and night was acquired to help answer these questions. The trail camera accounts for about 80% of the animal images presented here. The camera was always stationed on our property, even though it could see beyond. And no food was given to lure animals to the camera.

In early times, Native Americans had a close relation with animals since the animals lived among them and were a source of food and clothing and tools. The animals became a part of rituals and were the subject of many stories. These rituals and stories contain important lessons that teach us about sharing the world with all who live in it.

ACKNOWLEDGEMENTS

A special thanks to my wife Suzanne Barker who made significant contributions to this project. And thanks to Don Usner and Key Sanders for their valuable assistance.

CHICA

Hello! My name is Chica and I am a Rhodesian Ridgeback. I live with my family in the foothills of Santa Fe. I also live in the midst of a number of wild animals, many of which I do not know. Our neighborhood contains a few scattered houses, not many paved roads, very few fences, and no street lights. It is a safe place for animals to come and go.

We have seen some of the animals and our neighbors also report this. Mostly they are coyotes and deer. I can hear them at night and can smell their tracks when I am out. I am a hound and am excellent at sniffing out new and old visitors.

I can see some animals when I am inside.
Our house has lots of glass which goes from floor to ceiling.
When I hear an animal sound, I can check it out.

Here I had just returned from a walk and detected an animal scent.
Sniffing in the snow can be a bit challenging.

TRACKS AND TRAILS

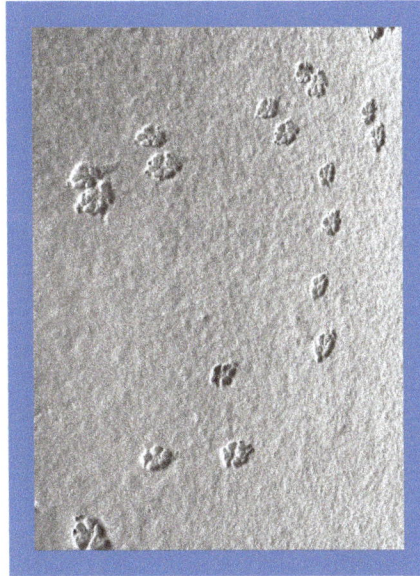

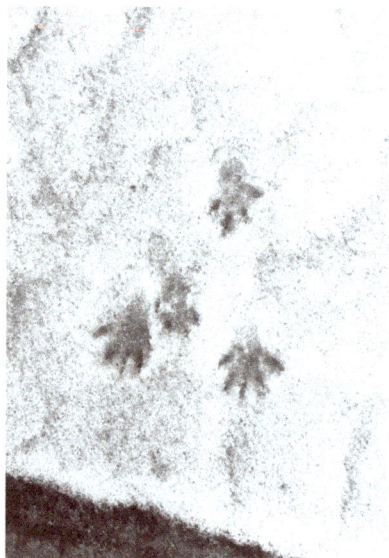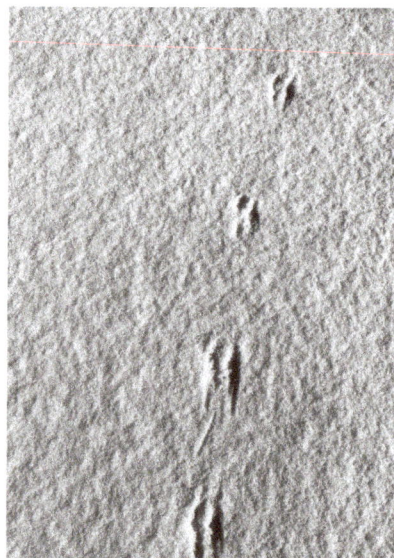

After a snow or rain, lots of tracks become visible. It is clear that these tracks were not made by coyotes and deer. Who are the animals that made these tracks? This is what we would like to know.

Many of the animals come from nearby Sun Mountain. There is an arroyo from the mountain to our neighborhood which is a highway for animals living around here to come and go. I know the arroyo because my family takes me for walks there. The arroyo is quite long and there are many points where the animals can enter.

Animals have been seen at almost every spot on our property. Some tend to follow trails which can be seen on an aerial photograph of our house. These trails appear as light marks on the ground. The most traveled trails are highlighted with a dashed black line.

"TC"

TC on Station

My family and I started a project to try and photograph some of the animals.

"TC" joined the team to help with the photography. TC is a trail camera that can be stationed at a particular location and can be moved from one point to another. TC can take images in the daylight or at night. When there is motion, TC becomes alert and takes pictures – usually three pictures taken one second apart. When taken during the day, the pictures are in color and at night they come out black and white. The pictures are made in all kinds of weather, including when it rains or snows.

The photography TC does can be a bit tricky. The animals do not pose. Often, they are moving rapidly and the shutter speed cannot be changed to compensate. They are seldom in a position that allows for a well framed image. Sometimes the image needs to be reframed or adjusted when the animal is far away.

To be good at this type of photography TC must know where the animals move about. TC must be stationed to get the best views, and many images must be taken. My job is to always be on the lookout to recommend where TC's location should be.

Animals create their own paths, and sometimes different animals take different paths that intersect. It is my responsibility to sniff these out and know where they are.

In doing the photography project, my family and I wanted to learn more about the animals; where they came from; how they interact with each other; and what did people who lived here in earlier times think of them. The Pueblos and Ancestral Pueblos before them have lived in the area for ten thousand years (the Spanish 500 years, the White Caucasian 200 years). There is a wealth of very interesting information, including many stories in folklore.

My family has told me several of these stories from the Native American people who lived here before us. Some of the stories, which are told again here, are marked by italics.

• • •

In Native lore, the earth is viewed as the mother of all things. All creatures and plants depend on earth for food, shelter, and water. If the earth is our mother, then all creatures are our sisters and brothers. Since many animals can see, hear, smell, and feel things that humans do not, it gives credence to the Native beliefs that much can be learned from watching, and listening to the animals.

OUR ANIMAL NEIGHBORS

COYOTE

The coyote is the centerpiece of our local wildlife, because the coyote connects to most of the other animals around here. The coyote hunts with the badger, socializes with the raccoon, is the major predator of young deer, and also the predator of the grey fox and the quail. The coyote coexists with the bobcat, and the coyote and crow monitor each other for sources of food.

The coyote is very popular character in Native stories. Usually, the coyote is a villain, almost never a hero with major power. The coyote can be a mixture of both good and evil, intelligent, cunning, clever, and almost always a sneaky trickster.

Coyotes often move their cubs from a den in one place to one in another place. They do this in order to protect their cubs from predators. TC was able to get pictures of this. One is in the daytime and the other is at night. In each image the mother carries the pup in her mouth.

• • •

In one Native story, a coyote saw a crow in a tree with a piece of deer fat in its beak. The coyote was hungry but could not climb the tree to get the meat. So, the coyote enticed the crow to sing. The food fell from the crow's beak and the coyote ate it.

After seeing these pictures, my family searched and found one of the dens after the pups had been moved. The den was not all they found; some of my missing toys were scattered nearby.

It is said that a coyote will eat anything it can chew. And I believe it. Here is a photo that TC took of a coyote that found one of my old bones.

• • •

In the beginning, when predator animals were created, the gods told them that they should fast for four days without food or water, and would then be given prey. Coyote and Wildcat got hungry and thirsty and took food on the third day. This broke the fast and from that time on, coyotes and wildcats had to search for food.

Coyotes in a group—watch out! Coyotes can be intimidating and dangerous around both pets and small children. I have seen as many as five on our property at one time and, even though I am bigger and stronger than a coyote. I keep my distance when there are several of them.

BADGER

The Badger is the scariest animal of them all! You don't see badgers around very often. They are ferocious diggers. They can dig faster than a person with a shovel. Wow! They hunt with coyotes. A coyote will chase a rabbit into a hole and the badger will dig it out at the other end of the tunnel. And if the badger is digging and the rabbit escapes out the other hole, the coyote will get it. Don't mess with badgers! Their claws are like knives!

• • •

The badger fetish is a symbol of strength, aggression, and healing. Badger uses strength, aggression and its superior digging ability to achieve its goals. As a digger it has knowledge of plants and roots used in healing. Native American folk lore associates the badger with the underworld and it is a symbol of warrior energy.

HAWK

In the past two years, TC has taken over 2000 images of animals. My favorite picture is of a hawk that is coming in for a landing. Hawks are very strong birds that can fly over 100 mph.

Birds are hard to photograph and flying birds are next to impossible. I plan to submit this hawk picture to a photography contest since I believe it is a sure winner!

In Pueblo lore the hawk and eagle are symbols of power.

• • •

A famous Pueblo story tells of a hawk and crow. The crow lays the eggs in a nest, becomes impatient waiting for the eggs to hatch, and abandons the nest because the eggs were taking too long to hatch. A hawk sees the eggs in the nest, sits patiently on the eggs until they hatch, and then feeds the baby crows and teaches them to fly. The crow returns and wants the baby crows but the hawk wants to be the mother of the baby crows she has raised. The eagle then has to determine who will be the mother. What do you think the eagle decides?

RACCOON

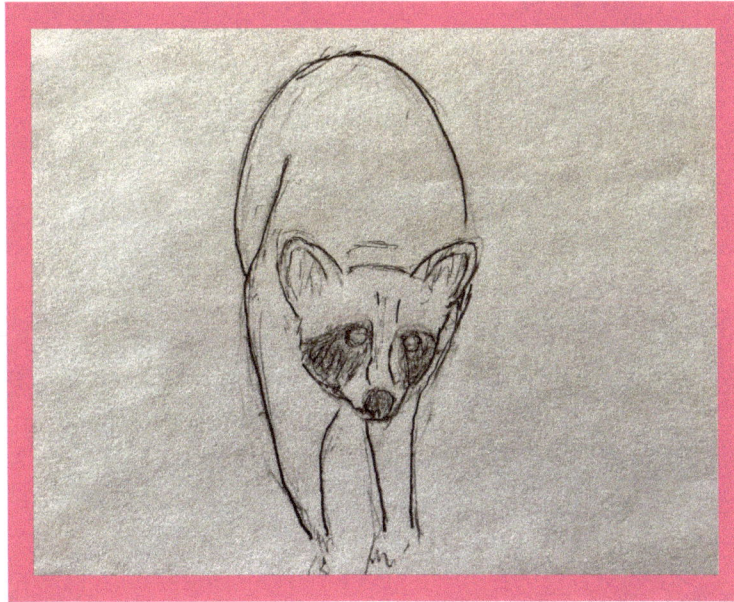

Next are the raccoons. Where is the water? Folklore tells us that raccoons require lots of water and even wash their food, but this turns out not to be the case. They do like to dip their paws in water because it sensitizes them for better feel.

Raccoons use their cleverness and dexterity to escape danger and acquire food. They are minor-league tricksters when compared to the coyote.

In terms of genealogy raccoons are not related to possums, or foxes, or similar animals. They are actually related to bears.

• • •

Once a raccoon saw a crow that had five very beautiful rings. The raccoon wanted them so badly that he decided to steal them. He put on a mask and looked for the rings. When he found them, he put them on his tail to escape. But he was caught. As punishment from then on, the raccoon had a mask on its face and rings on its tail.

BULL SNAKE

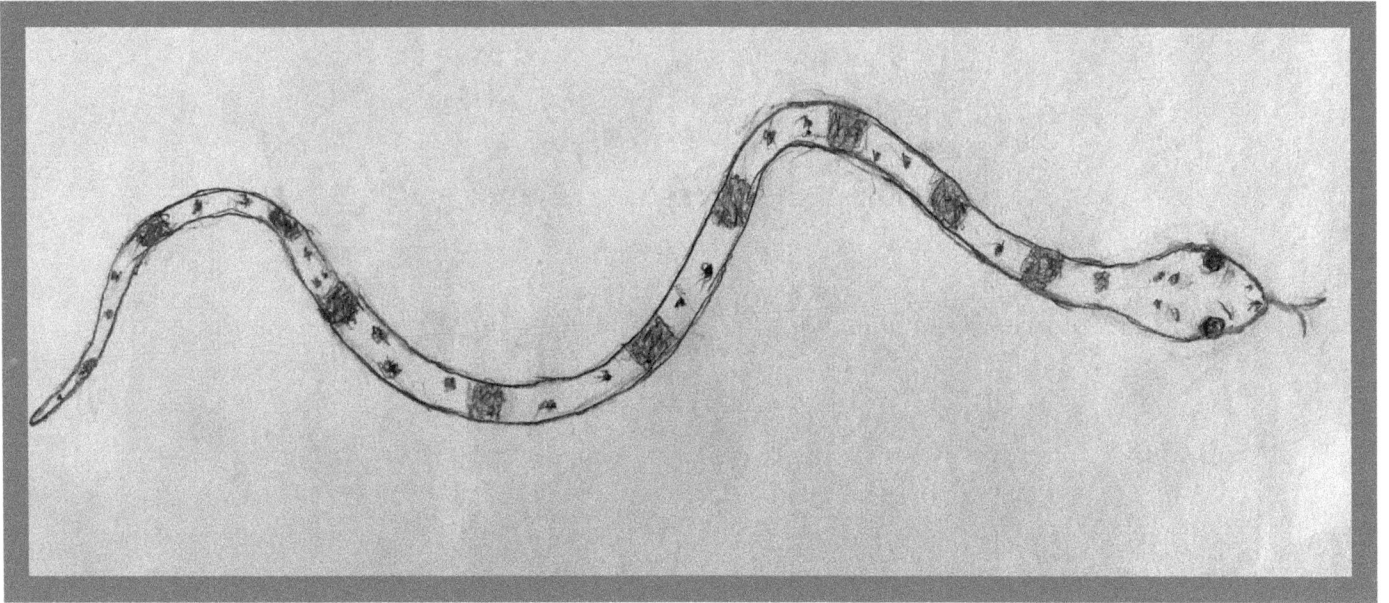

Slither moving along the edge of the wall.

Slither coming and going out of the rock wall.

We happened to find a huge bull snake (six feet long) living in the rock wall behind our house. We named it Slither. Slither would join us from time to time and move about when we were sitting near the wall.

Slither's head and tail

DEER

Large numbers of deer pass TC's stations. They come single or in groups of between two and seven. In terms of photos, they are third in number only to the coyotes and rabbits. As a result, TC has been able to get many images of deer.

Every spring male deer engage in combat to see who is number one. Their battles are more like a boxing match than a fight. One spring I had a ringside seat and saw two bucks locking horns and pushing hard against each other right in front of me. Other deer spectators were watching, too.

When that contest was over, one buck had lost one of his antlers. I then chased the other deer spectators and they ran away. The sound and rhythm of antlers clashing is thought to be motivation for the Pueblo Deer Dance.

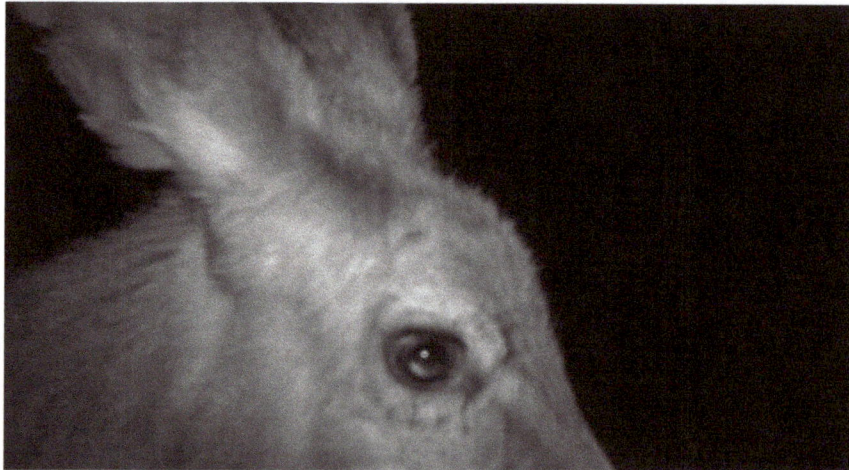

I can tell you this. The deer are really intrigued by the camera and like to be photographed. They often just stick their faces right up to TC's camera lens. It doesn't matter if it is day or night. In fact, one night a buck with one antler came by and TC took 170 images!

Here are 21 of the 170 images. Does this make my case?

BOBCAT

Here is a bobcat crouching and stalking a prey!

Native American tales explain that bobcats are loners. The lesson the bobcats teach is about being alone without being lonely. Very serious animals, bobcats are almost always out at night and almost always alone. If you see a bobcat, it is okay. Just don't try to make it a pet!

GRAY FOX

Of all the animals that have been photographed, a gray fox I call Foxy is my favorite. Foxy is an unusual fox. The Gray Fox came from the canidae species about three thousand years ago. This makes it a relative of the dog and the wolf, at least that's what I have heard. I know when I see one because it has a large bushy tail with a black stripe down the middle.

Gray foxes are quite small, weighing less than 10 pounds. There is evidence that in earlier times, people tried to domesticate them, but the gray fox wants to be free in the wilds. They eat small mammals such as mice, rats, and rabbits. Their enemies are coyotes and bobcats. The gray fox is the only fox that can climb and live in a tree. They are not urban animals and are very elusive and rare to spot in the wild.

CROW

In Native culture and mythology the crow symbolizes the highest form of intelligence and wisdom. Many believe crows have the ability to communicate with humans. Examples exist today where crows bring gifts to humans. My family keeps a water dish for birds. The crows recognize my family and make a tremendous fuss when the water dish is empty. They can recognize and remember a human face as a friend or foe for several years. It is not surprising that they are popular characters in the ancient stories.

Crows are the "eyes and ears" of the neighborhood. When they are circling and squawking, something unusual is going on. They definitely interact with and alert the community of animals when something is happening.

Crows are not friends with hawks. The hawks raid crow nests to take eggs and chicks. When a hawk is flying in the area, several crows start chasing the hawk and take turns dive-bombing the hawk until the hawk flies far away.

QUAIL

Can you find all six quail in the above picture?

I see a lot of quail, mostly in groups. I have seen one pair around here for at least three years.

The quail family of two quail that lived here has grown to a covey of six. They stay mostly under bush and are speedy little birds, especially when they cross open areas. They rest at night in a circle with their heads pointing out so they can explode in all directions when there is danger.

• • •

Quail are viewed by Natives as humble birds and are sometimes associated with earth. They teach us to be aware of everything around us, to be mindful of danger, and to react quickly in times of crisis.

RABBIT

Rabbits are the most prolific of the animal photos. TC and I agree that rabbits do not make very good trail camera images. The rabbits are physically the smallest animals that are photographed (except for the occasional rodent), and appear even smaller unless they are close to the camera. The pictures are often a bit out of focus unless the rabbit is sitting still when the image is taken.

Rabbits are extremely important since they are prey for all carnivores that live here or pass through. In a sense, they are the engine that drives the local animal economy.

• • •

For Native Americans, rabbits are a major source of food and clothing. When the Pueblo people speak of hunting without naming the animal, it is understood they mean hunting for rabbits. Rabbits have few defense mechanisms, the strongest being the ability to reproduce. There always seems to be enough rabbits to fit the demand for food. The result is a strong link between rabbit and man. An important lesson that rabbit teaches us is that while the body is transitory, spirit is immortal.

ON THE TRAIL AGAIN

The animals in these photographs live and move among us, but they aren't the only animals that live here. There are more. I heard that several years ago a mountain lion was spotted as well as a bear.

Helping TC photograph the animals—and even people when they come into view—is loads of fun. I love to discover when a new species has come through and follow its path.

I continue to patrol the property and when I detect the smell of a new animal, I identify the coordinates. Then TC will be stationed there and ready to photograph.

READERS GUIDE

1. Why do you think Chica was selected to be narrator of this story?

2. Did TC take photographs of animals you did not expect to see in the foothills of Santa Fe? Are there other animals you would expect to see?

3. Did the animal images and Native stories affect your view of the animals as neighbors?

4. Chica's favorite animal is the gray fox. Which is yours and why?

5. When it snows or rains, do you look for animal tracks around you? Can you identify them?

6. Animals need to find food to survive. Can you diagram these animals according to who is predator and who is prey?

7. In most Native stories, coyote is usually a villain or prankster. Yet in some stories coyote is involved in the creation of the world. Can you explain this apparent inconsistency?

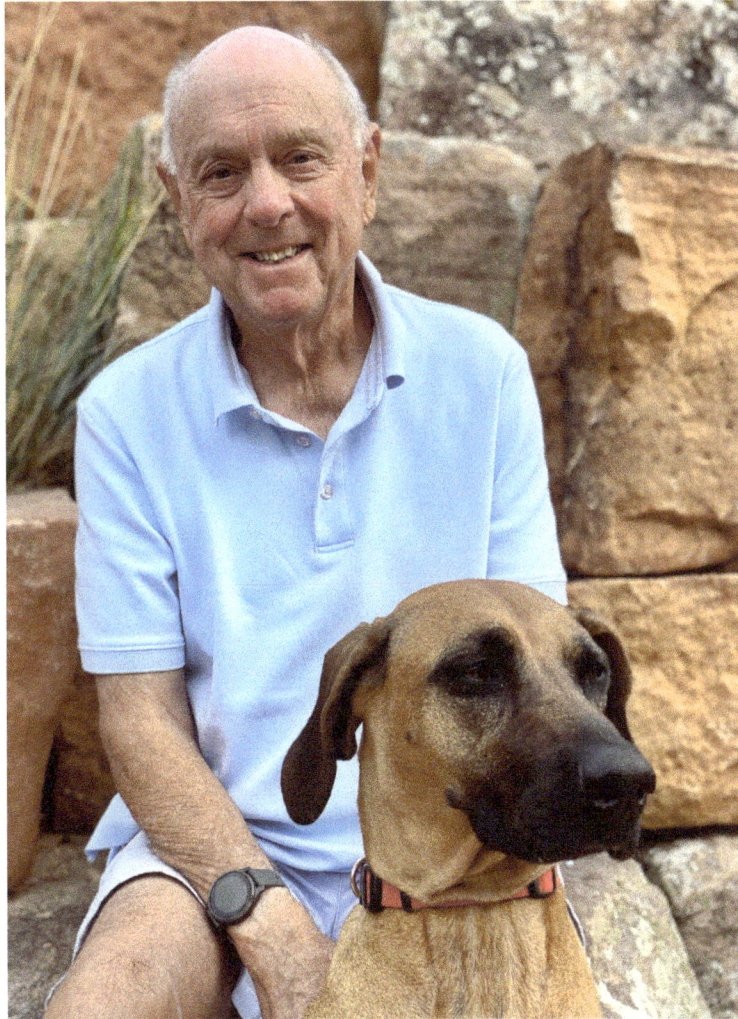

www.ingramcontent.com/pod-product-compliance
Lightning Source LLC
Chambersburg PA
CBHW050048220326
41599CB00045B/7330